DIAGRAMS
OF THE NERVES
OF THE HUMAN BODY.

FINE ILLUSTRATED WORKS

GODLEE An Atlas of Human Anatomy, illustrating most of the Ordinary Dissections, and many not usually practiced by the Student, accompanied by an Explanatory Text By RICKMAN JOHN GODLEE, M S F R C S, Fellow of University College, Assistant Surgeon to University College Hospital, and Senior Demonstrator of Anatomy in University College
In Two Volumes (folio Atlas and octavo volume of Text), Fine Cloth, $30 00
Or, in 12 Parts, each with accompanying Pamphlet of Text, per Part, $2 50
This series of Plates has received the unqualified approbation of eminent members of the profession, and reviewers have freely accorded it praise
Sold only by Subscription

MARTIN Martin's Atlas of Obstetrics and Gynæcology Translated and edited from the Second German Edition with additions, by FANCOURT BARNES, M D, M R C P, London
One Volume, 4to, bound in Heavy Beveled Boards, $12 00
" A book of the highest class, and one with which any obstetric physician may be proud to connect his name '
" Of great service to practitioners and students '
Sold only by Subscription

BENTLEY AND TRIMEN Medicinal Plants Being Descriptions, with Original Figures of the Principal Plants employed in Medicine and an account of their Properties and Uses Complete in Forty-two Parts Price $2 00 each With 300 Colored Illustrations (natural size) Large 8vo In 4 volumes Half Morocco and Gilt Top Price $90 00
This work is an Illustrated Botanical Guide to the British, United States and Indian Pharmacopœias, it includes also other species employed, or in common use, though not officinal

HEATH Operative Surgery A Course of Operative Surgery, consisting of a Series of Plates, each plate containing Numerous Figures, Drawn from Nature, Engraved on Steel, and Colored by Hand, with Descriptive Text of each Operation By CHRISTOPHER HEATH, F R C S Complete in Five Quarto Parts, each containing Four Large Plates and numerous Figures
Price per Part, $2 50, or bound in one vol Cloth, $14 00

HUTCHINSON Illustrations of Clinical Surgery Consisting of Plates, Photographs, Wood cuts Diagrams, etc, illustrating Surgical Diseases, Symptoms, and Accidents, also Operations and other Methods of Treatment With Descriptive Letter-press By JONATHAN HUTCHINSON F R C S Quarterly Fasciculi Imperial Quarto *Parts One to Fourteen Now Ready*, $2 50 each, or the first ten parts bound in one volume, complete in itself, Price $25 00

FOX Atlas of Skin Diseases Complete in Eighteen Parts, each containing Four Chromolithographic Plates, with Descriptive Text and Notes upon Treatment By TILBURY FOX, M D, F R C P, Physician to the Department for Skin Diseases in the University College Hospital
Folio size Price $2 00 each, or complete, bound in Cloth, Price $30 00

BRAUNE Atlas of Topographical Anatomy Containing Thirty-four Full page Photolithographic Plates after Plane Sections of Frozen Bodies, and Forty-six Large Wood Engravings By WILHELM BRAUNE, Professor of Anatomy in the University of Leipzig With Marginal References, each Plate accompanied with Full Explanatory Text Translated by EDWARD BELLAMY, F R C S A large imperial octavo volume
Price, bound in Cloth, $12 00, Half Morocco, gilt top, $14 00

JONES Aural Atlas An Atlas of Diseases of the Membrana Tympani With Sixty three Colored Figures, and appropriate Letter-press By H MACNAUGHTON JONES, M D Quarto, bound in Cloth Price $6 00

FRORIEPI Atlas Anatomicus Containing Thirty Large Plates, in all Seventy six Figures, with Full References to the Muscles, Arteries, Ligaments, etc By ROBERTI FRORIEPI, M D One Volume, Quarto Plain Plates, Price, $5 00 Colored Plates, Price $10 00

PRESLEY BLAKISTON, Publisher,
1012 WALNUT ST, PHILADELPHIA, PA

DIAGRAMS

OF THE

NERVES

OF THE HUMAN BODY,

EXHIBITING THEIR ORIGIN, DIVISIONS AND CONNECTIONS, WITH THEIR
DISTRIBUTIONS TO THE VARIOUS REGIONS OF THE CUTANEOUS
SURFACE AND TO ALL THE MUSCLES

BY

WILLIAM HENRY FLOWER, F.R.S.

THIRD EDITION

PHILADELPHIA
PRESLEY BLAKISTON,
1012 WALNUT STREET
1881

ILLUSTRATIONS.

PLATE I   THE CRANIAL NERVES

1 The Olfactory Nerve   2 The Optic Nerve   3 The Third Nerve (*Motor Oculi*)   4 The Fourth Nerve (*Pathetic* or *Trochlear*)   5 The Fifth Nerve (*Trifacial* or *Trigeminal*)   6 The Sixth Nerve (*Abducens Oculi*)

PLATE II   THE CRANIAL NERVES

1 (VII) The Facial Nerve, or *Portio Dura* of the Seventh Pair   2 (VIII) The Auditory Nerve *Portio Mollis* of the Seventh Pair   3 (IX) The Glosso-Pharyngeal Nerve   4 (X) The Pneumogastric or Vagus Nerve—the *Par Vagum* of the Eighth Pair   5 (XI) The Spinal-Accessory Nerve   6 (XII) The Hypoglossal Nerve

PLATE III   THE SPINAL NERVES

The distribution of the Cervical and Dorsal Nerves is shown

PLATE IV   THE SPINAL NERVES

The remaining Spinal Nerves, including the Lumbar Plexus and the Sacral Plexus

PLATE V   THE SYMPATHETIC SYSTEM OF NERVES

The Trisplanchnic, Ganglionic, or Nervous System of Organic Life   1 Cervical Ganglia   2 Thoracic Ganglia   3 Lumbar Ganglia   4 Sacral Ganglia   5 Coccygeal, or Ganglion Impar.

PLATE VI   THE DISTRIBUTION OF THE CUTANEOUS NERVES

Showing the sources from which the sensibility of the different regions of the Cutaneous Surface is derived

PREFACE.

These diagrams were originally published in 1860. They were designed by the author while engaged in teaching anatomy in the medical school attached to the Middlesex Hospital. They have no pretensions to enlarge the boundaries of the knowledge already existing upon the subject to which they relate, but aim only at placing that knowledge in a form easily accessible, as well to students as to those whose avocations no longer afford the time or opportunity for anatomical investigations.

The distribution of all the nerves of the body, so far as the branches have received distinctive appellations, is shown, and their divisions are traced to the muscles, and to the various regions of the cutaneous surface. To afford greater facility for reference, the names of the muscles are printed in red letters, those of the nerves being in black.

It must be clearly understood that the plates are only diagrams or plans, and that in reducing to a plane surface objects which are in reality superimposed at various distances, and which sometimes cross one another, their mutual relations and proportions must often be disarranged.

In the plexuses and other parts which vary somewhat in different subjects, the average distribution in in its most simple form has been selected for illustration, and in difficult or disputed points, such as the connecting branches of the cranial nerves, only those which are established on good authority are introduced.

As few things tend to embarrass the student so much as a diversity of nomenclature, that used in the sixth edition of Quain's *Elements of Anatomy*, by Dr Sharpey and Mr Ellis, has been adopted throughout

The principal materials for the composition of these diagrams have been obtained by repeated dissections, but the author desires to acknowledge the assistance derived from the above-named work, from that of Swan, and from the beautiful plates of Hirschfeld and Leveillé

In the second edition, published in 1872, the size of the diagrams was considerably reduced, the convenience and portability of the work being thereby increased, without any detriment to its clearness or legibility In revising the plates, the author availed himself of some suggestions for which he was indebted to the kindness of Professor Turner, of Edinburgh, and of Mr J Beswick Perrin Some additional slight emendations have been made in the present edition

ROYAL COLLEGE OF SURGEONS OF ENGLAND,
December, 188c

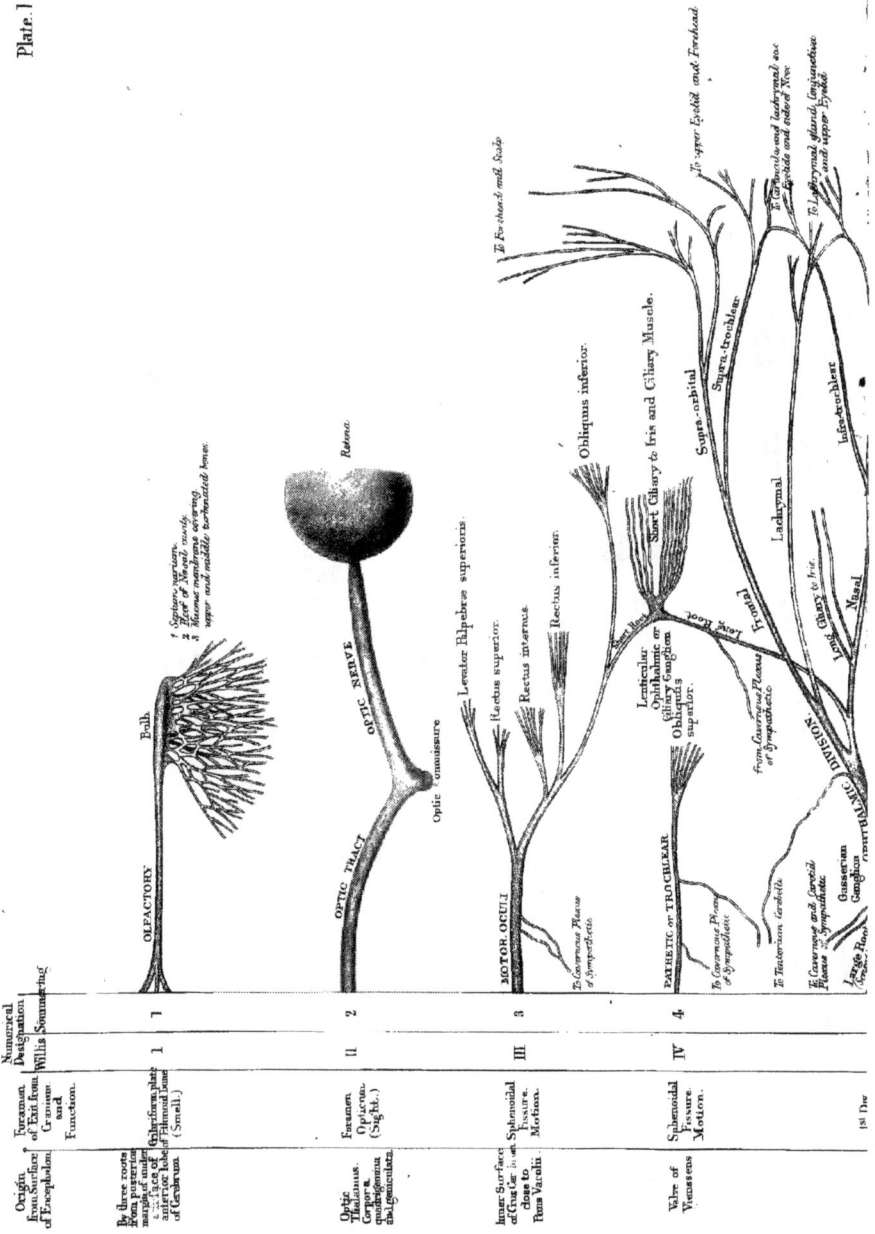

Plate 1.

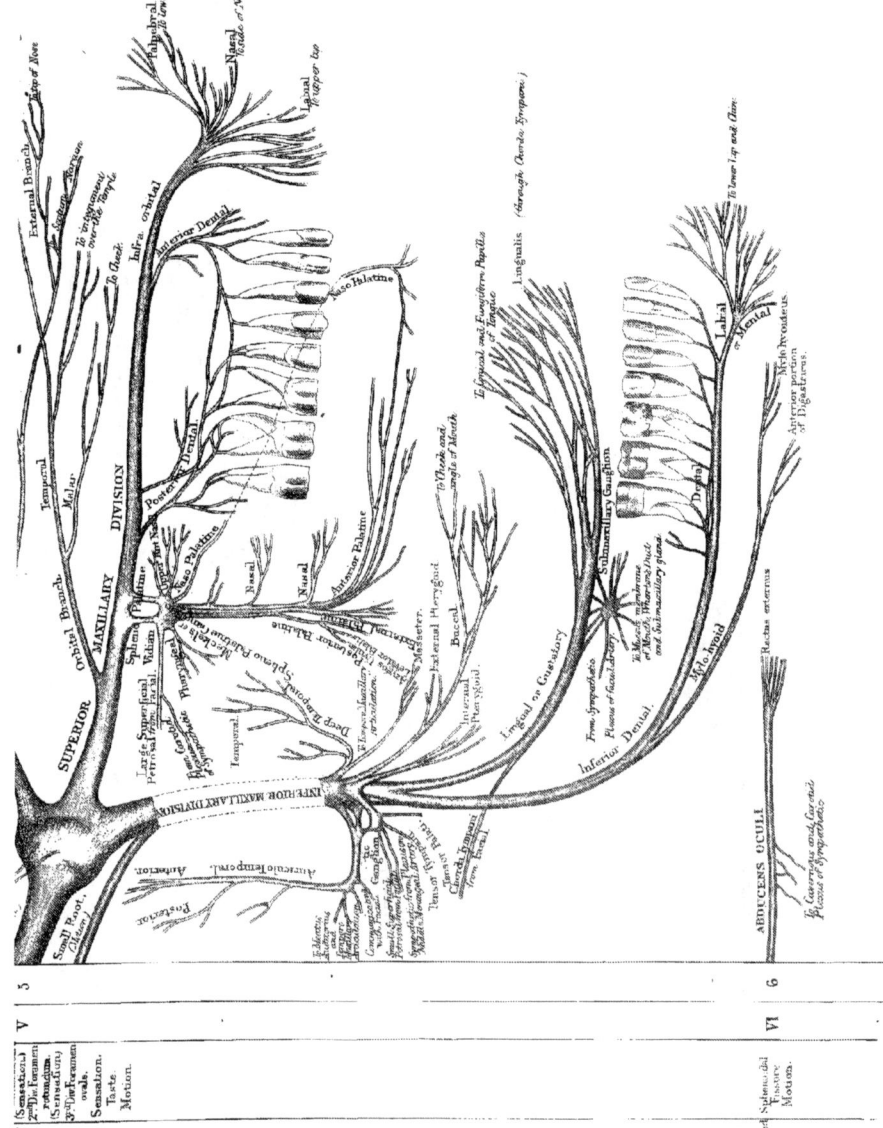

Plate ii.

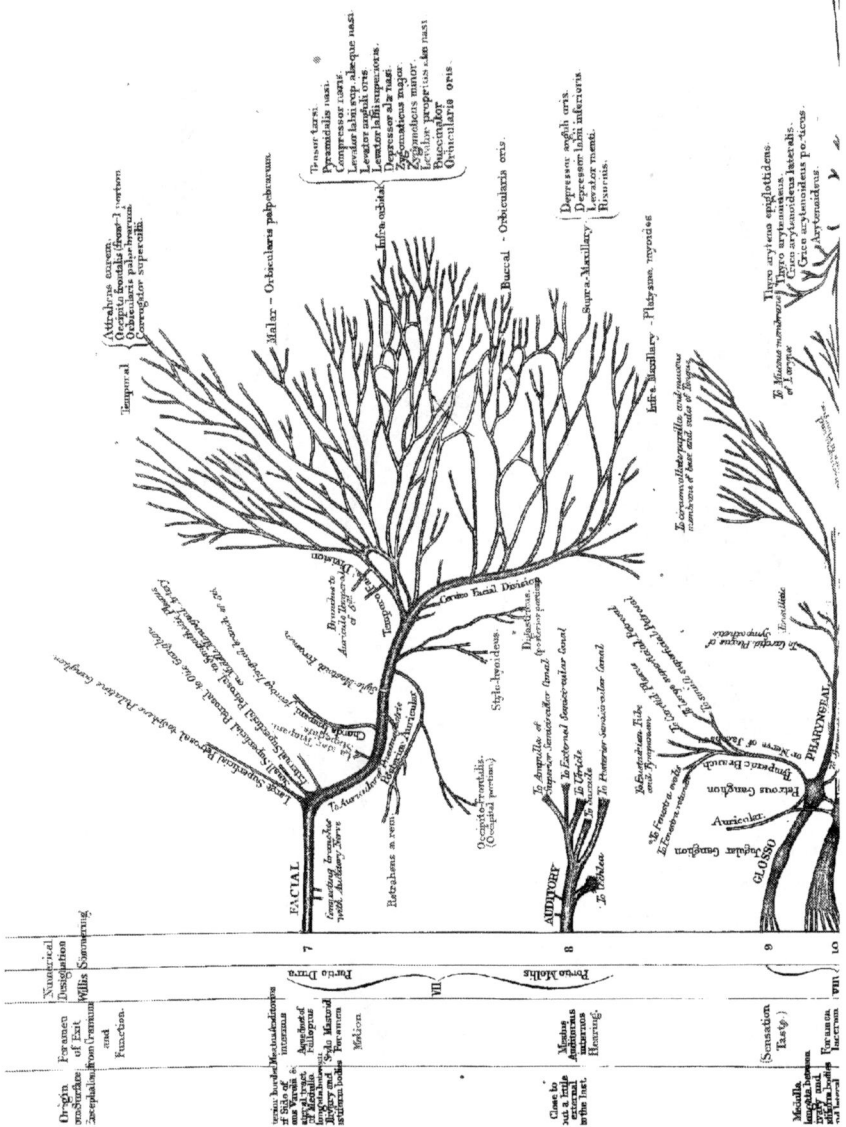

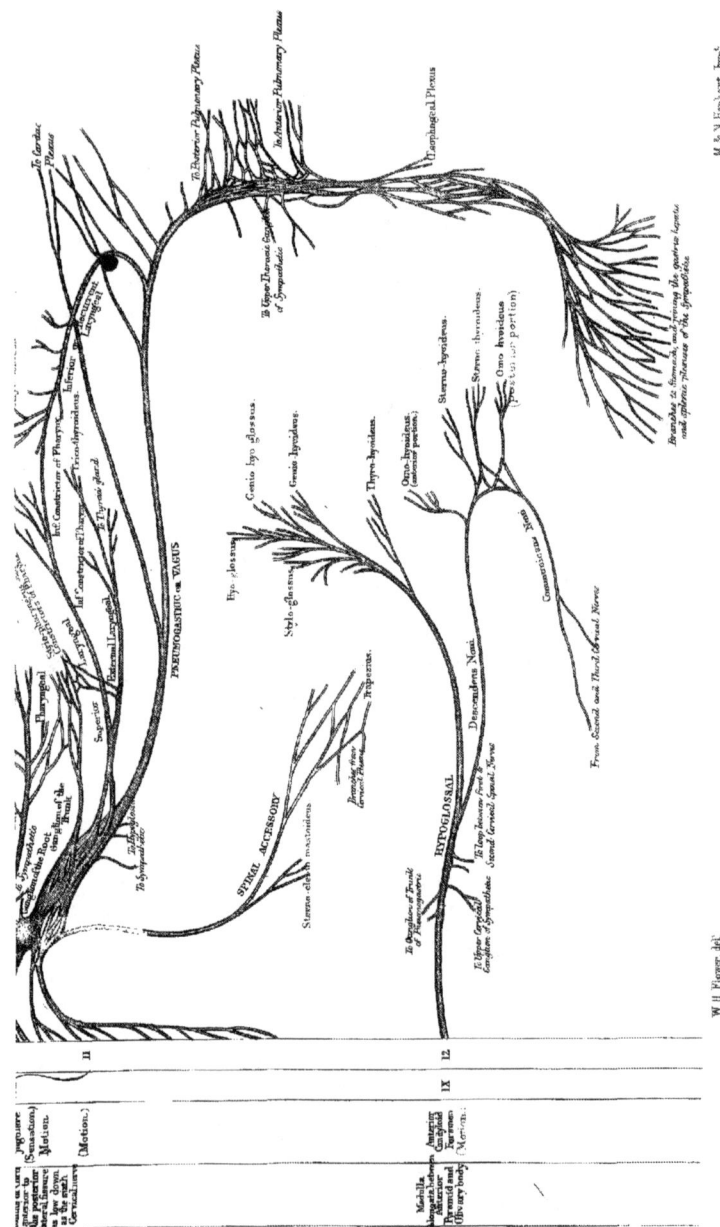

EXPLANATION OF THE PLATES.

PLATES I AND II.
THE CRANIAL NERVES

UNDER this name are included the twelve pairs of nerves arising from the cerebro-spinal axis, which are transmitted to their destination through apertures in the base of the skull

The numerical nomenclature of the cranial nerves, introduced by Willis, had become so completely incorporated into our medical, anatomical and physiological literature before its errors were discovered, that it would be impossible now to discard it altogether. The arrangement of Sommerring, in which each of the five pairs of nerves forming the seventh and eighth of Willis are recognized as distinct, is more correct, and is already very generally used upon the Continent. In the diagram the two systems are placed for comparison in contiguous columns

In the first explanatory column the superficial or apparent origins of the nerves only are given Although the deep or real origins are unquestionably of greater physiological importance, they are so complex and in many cases still so incompletely made out, that they could not have been accurately stated in that concise form required in a tabular exposition

Most of the cranial nerves are connected by fine filaments with branches of the sympathetic system The physiological import of these communications being still imperfectly understood, the words *to* and *from* applied to them must be taken in an anatomical sense only, and not necessarily as implying the presumed direction of the current of nerve force

The first plate illustrates the distribution of the first six pairs of cranial nerves

I THE OLFACTORY NERVE is specially appropriated to the sense of smell Although commonly described as a nerve, this with its expanded extremity should more properly be considered as a portion of the encephalon, as it contains much gray matter, has no sheath or neurilemma, and is homologous to the more fully developed *olfactory lobes* of the lower animals

From the under surface of the bulb about twenty delicate nerves are given off, these pass through the foramina in the cribriform plate of the ethmoid bone, and are distributed to the mucous membrane lining the upper two-thirds of the nasal fossæ

II THE OPTIC NERVE is the special nerve of the sense of sight, and terminates in the retina. The nerves of opposite sides are connected together at the *commissure*. From their origin up to this point, being in the form of flattened bands, they are called *optic tracts*.

III THE THIRD NERVE (MOTOR OCULI) is entirely motor in its function. It supplies branches to five of the seven muscles of the orbit, and one (the short root) to the ophthalmic ganglion, thereby giving motor power to the iris.

IV THE FOURTH NERVE (PATHETIC OR TROCHLEAR) is the smallest of the cranial nerves, and terminates in the superior oblique muscle of the eyeball. It communicates with the sympathetic, and often with the ophthalmic division of the fifth, and according to Bidder, whose observation has been confirmed by Hirschfeld, gives a recurrent filament to the dura mater.

V THE FIFTH NERVE (TRIFACIAL OR TRIGEMINAL) resembles the spinal nerves in having two roots, one endowed with sensory, the other with motor power, but the similarity is only partially carried out, inasmuch as the motor root is of very small size and of comparatively limited distribution.

It is the principal nerve of common and muscular sensibility to the face, confers motor power on the muscles of mastication, and one of its branches contains filaments appropriated to the special sense of taste. As in the spinal nerves, the sensory root has a gangliform enlargement upon it (the Casserian ganglion). From this proceed the three main branches or divisions of the nerve. The two upper divisions are purely nerves of common sensation, while the lower, into which alone the motor root passes, is of very complex character, having among its branches—

1. Simple nerves of sensation (inferior dental, auriculo-temporal, buccal).
2. Nerves which convey motor power to certain muscles, and probably, like the spinal nerves, contain sensory fibres also.
3. A nerve of ordinary sensation and special sense combined (gustatory), and which by the addition of the chorda tympani from the seventh nerve contains motor filaments.

The branches of the fifth nerve towards their termination communicate freely with those of the seventh, and confer sensibility upon the muscles of the face, which receive their motor power from the latter.

In connection with this nerve are four small ganglionic masses, containing gray matter, each of these appears to be in communication (by *roots*) with a motor, a sensory, and a sympathetic nerve, and to give *branches* of distribution to contiguous structures.

VI THE SIXTH NERVE (ABDUCENS OCULI) is the motor nerve of the external rectus muscle of the eyeball.

The second plate shows the distribution of the remaining six pairs of cranial nerves.

VII THE FACIAL NERVE, or *Portio dura* of the seventh pair (Willis), is purely a nerve of motion, and is distributed to the muscles of the face. Besides those named in the diagram in direct connection with it the seventh appears also to supply through the large superficial petrosal (which after being joined by a branch from the sympathetic, receives the name of Vidian, and enters Meckel's ganglion) some of the muscles of the soft palate, and through the chorda tympani the intrinsic muscular fibres of the tongue. Before their termination in the muscles, its branches communicate freely with the sensory fibres of the fifth nerve.

VIII. THE AUDITORY NERVE, or *Portio mollis* of the seventh pair (Willis), is the special nerve of the sense of hearing, and is distributed to the internal ear

IX. THE GLOSSO-PHARYNGEAL NERVE gives filaments through its tympanic branch to some parts of the middle ear, but it is chiefly distributed to the mucous membrane lining the upper part of the pharynx, the Eustachian tube, the arches of the palate, the tonsils, and to the sides of the posterior part of the upper surface of the tongue It is a nerve of the special sense of taste, and of ordinary sensation to the parts which it supplies, and is the chief centripetal nerve engaged in the action of deglutition It is doubtful whether it contains any motor filaments which are not derived from its communication with other nerves

X. THE PNEUMOGASTRIC OR VAGUS NERVE, the *Par vagum* of the eighth pair (Willis), has a most extensive distribution, giving branches to the pharynx, larynx, trachea, lungs, heart, œsophagus, and stomach Its main trunk being of great length, it has been necessary in the diagram to give it a curve so as to adapt it to the size of the paper It has numerous communications with other nerves, both cranial, spinal, and sympathetic, and its functions appear to be of very mixed character, partly motor, partly sensitive, and partly of a nature allied to those of the nerves of the sympathetic system

XI. THE SPINAL-ACCESSORY NERVE is apparently entirely motor in its function It arises from the upper part of the spinal cord, a considerable portion of it joins the pneumogastric (whence its name, "nervus *spinalis* ad par vagum *accessorius*"), the remainder is distributed to the sterno-cleido-mastoideus and trapezius muscles

XII. THE HYPOGLOSSAL NERVE The ninth pair in the system of Willis supplies all the extrinsic muscles of the tongue, as well as certain others in connection with the hyoid bone Its proper function appears to be exclusively motor, such sensibility as it possesses being probably derived through its free communication with the spinal nerves

PLATES III AND IV.
THE SPINAL NERVES

THE nerves which arise from the spinal cord have each two roots   Of these the posterior is somewhat larger than the other, has a ganglion situated upon it, and is composed solely of filaments which convey sensory impressions toward the cerebro-spinal centre   The anterior root, which has no ganglion, consists on the other hand of fibres which transmit motor power from the centre to the muscles   After the union of these roots, the resultant nerve is of mixed function, containing both motor and sensory fibres

Directly the nerves issue from the intervertebral foramina they divide into two branches, one of which, comparatively small, is directed posteriorly, and supplies the skin and muscles of the back   The anterior branches form the large nerves which are distributed to the neck, the lateral and anterior parts of the trunk, and the extremities

The spinal, like the cranial nerves, are symmetrically disposed on the two sides of the body   There are thirty-one pairs, divided for the convenience of description, as follows   cervical, eight, dorsal, twelve, lumbar, five, sacral, five, coccygeal, one

In Plate III the distribution of the cervical and dorsal nerves is shown

The anterior branches of the upper four constitute the CERVICAL PLEXUS   those of the four lower cervical, together with a large branch from the first dorsal, form the BRACHIAL PLEXUS   The anterior branches of the dorsal nerves are called intercostal   Those below the third have not been represented in the diagram, as they all resemble each other, running forwards between the ribs, supplying the intercostal muscles, and giving off lateral and anterior cutaneous nerves to the surface of the chest

Small filaments which pass from the anterior branches of all the spinal nerves near their commencement to the ganglia of the sympathetic system have been omitted here, to avoid the risk of obscuring any portions of the special objects of the diagram, but they will be seen in Plate V

Plate IV illustrates the distribution of the remaining spinal nerves   The anterior branches of the first three lumbar nerves and the greater part of that of the fourth constitute the LUMBAR PLEXUS   The large nervous cord formed by part of the fourth and the whole of the fifth lumbar, together with the first three and part of the fourth sacral nerves (anterior branches), is called the SACRAL PLEXUS

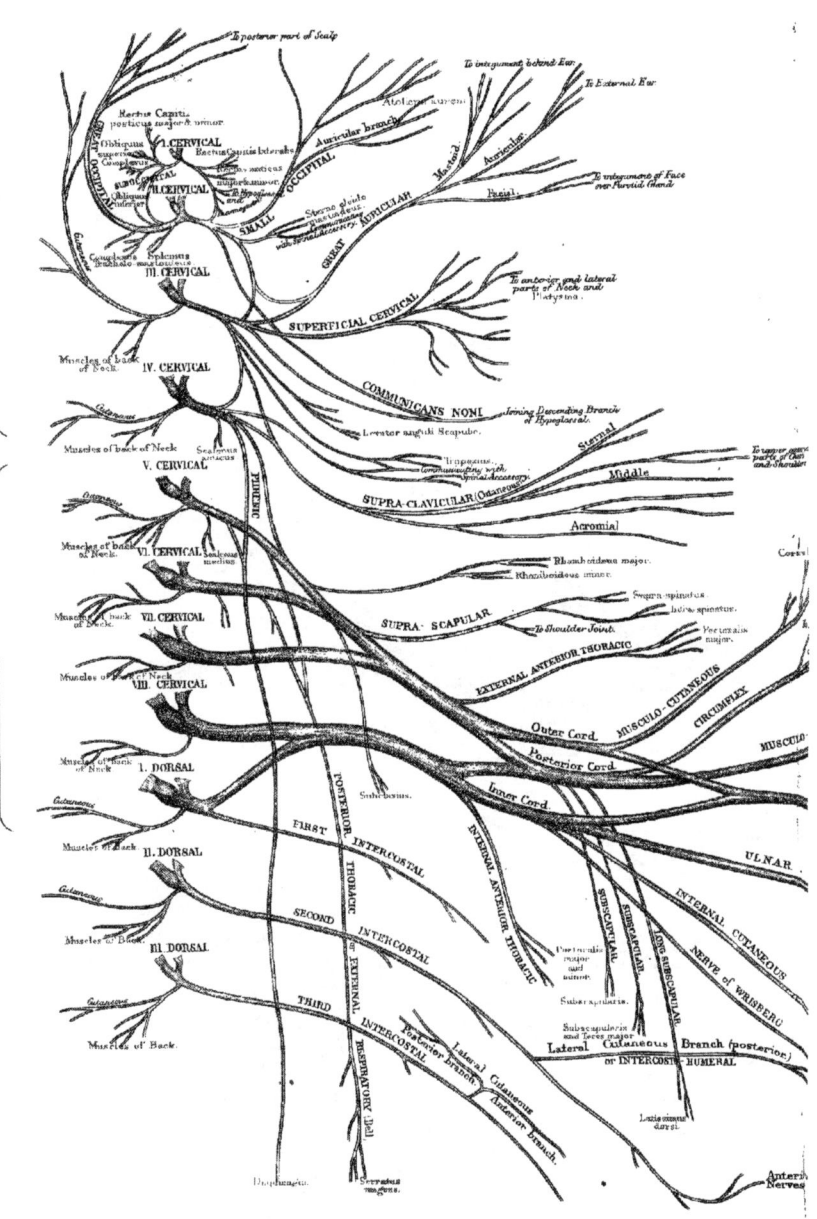

Plate III.

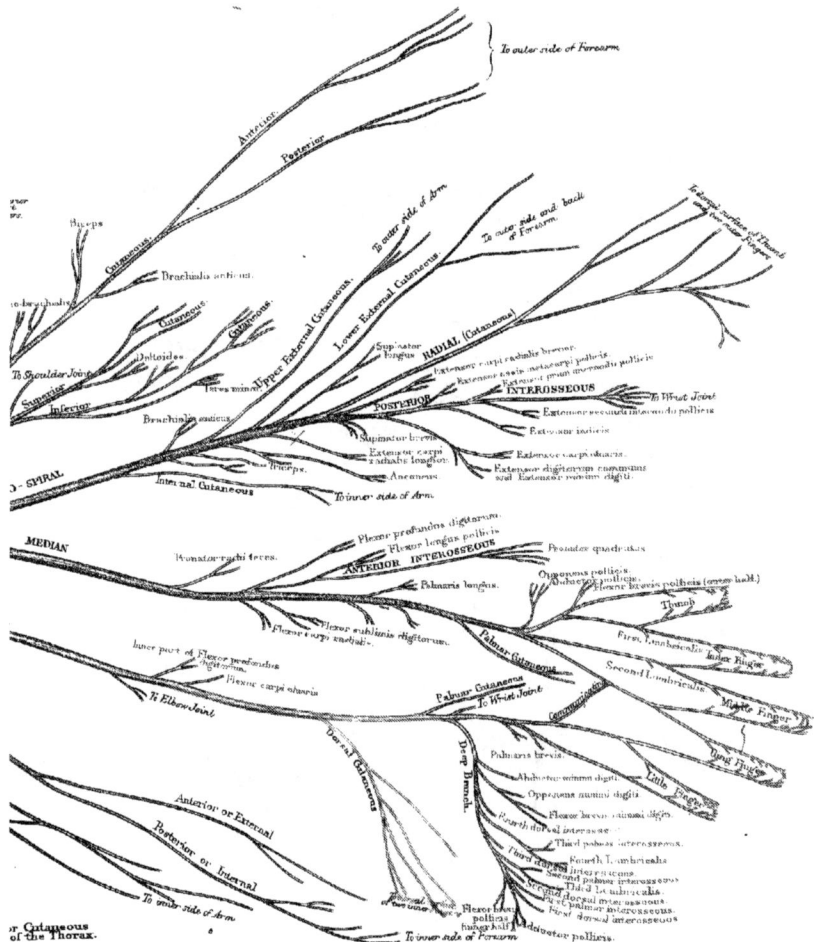

M & N Hanhart imp

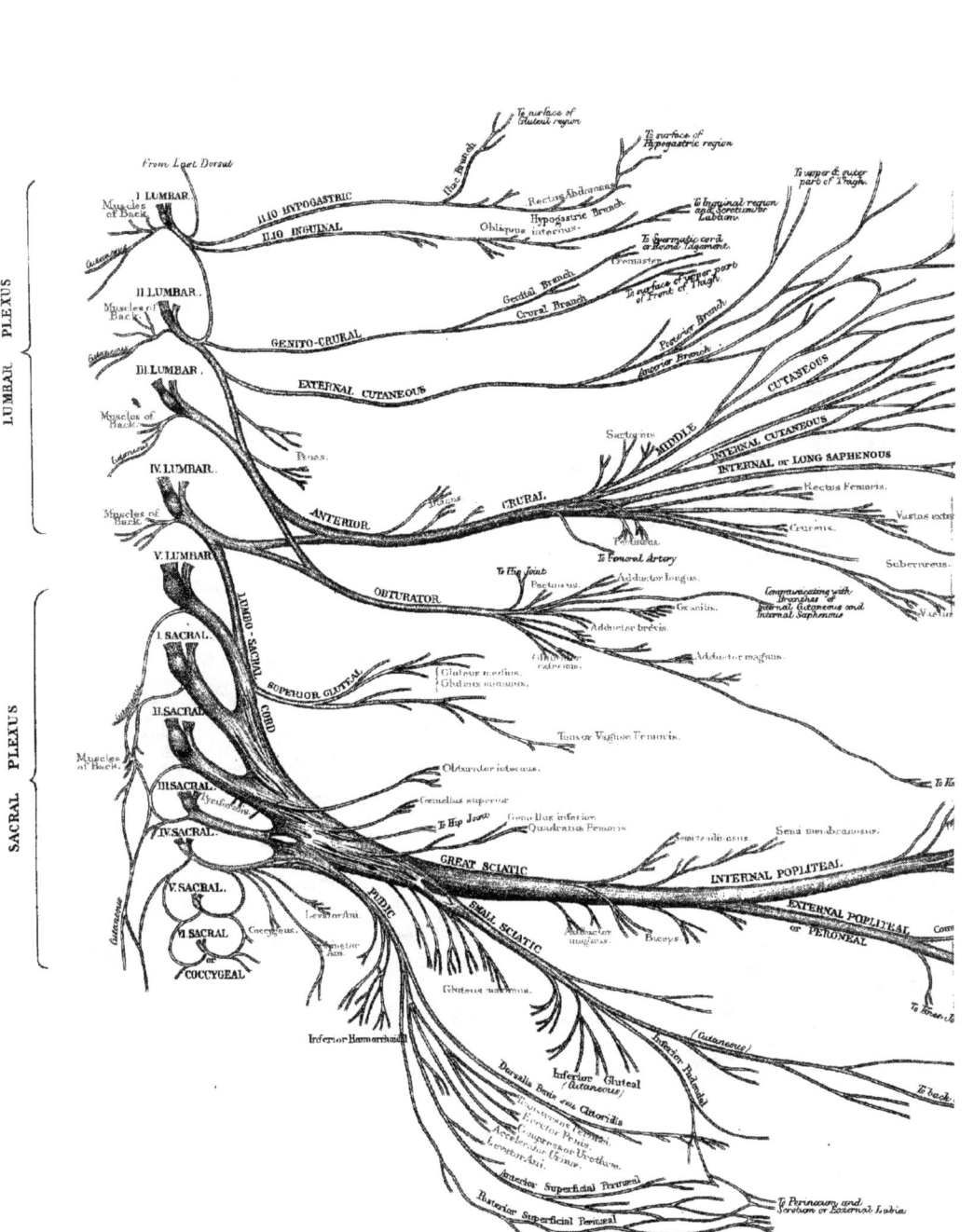

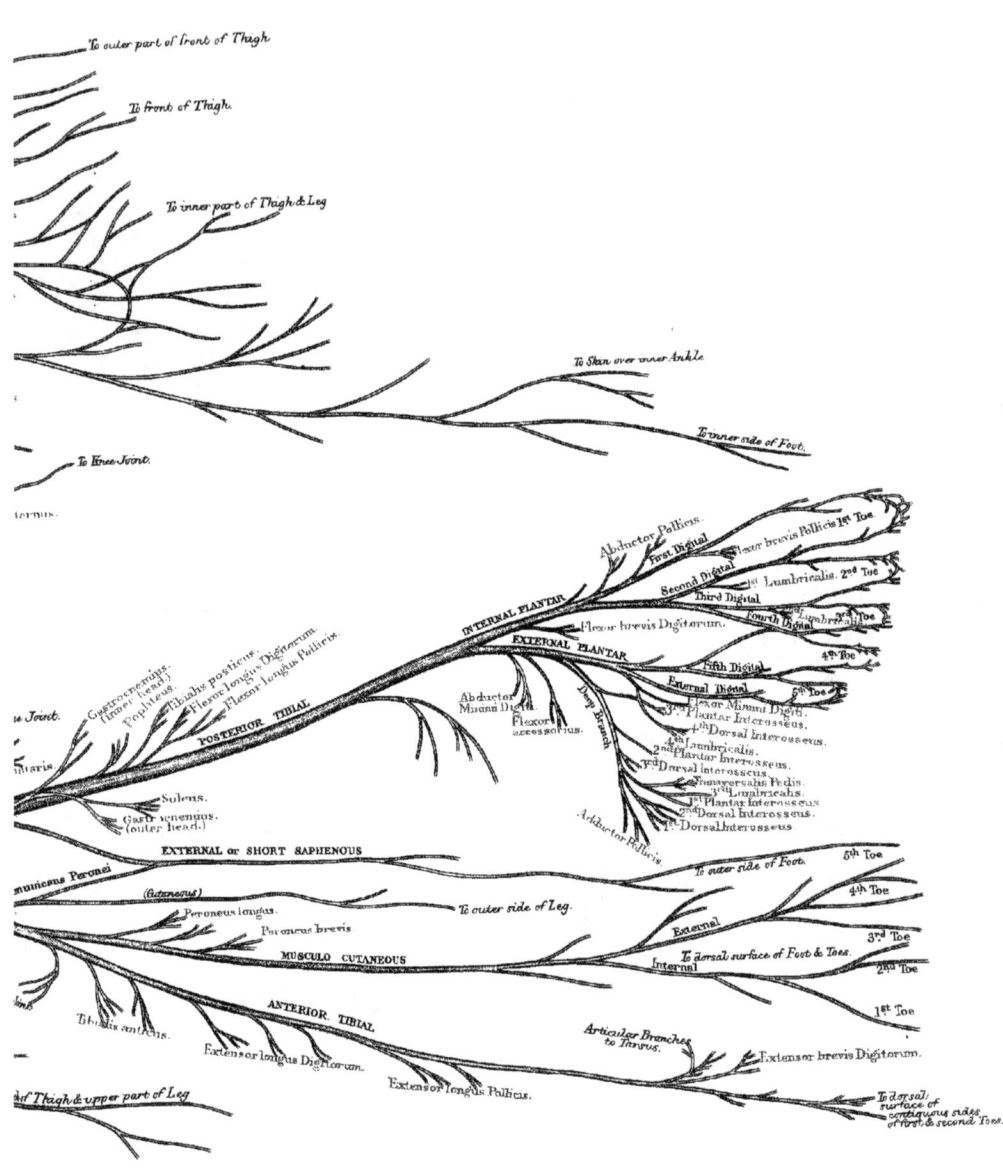

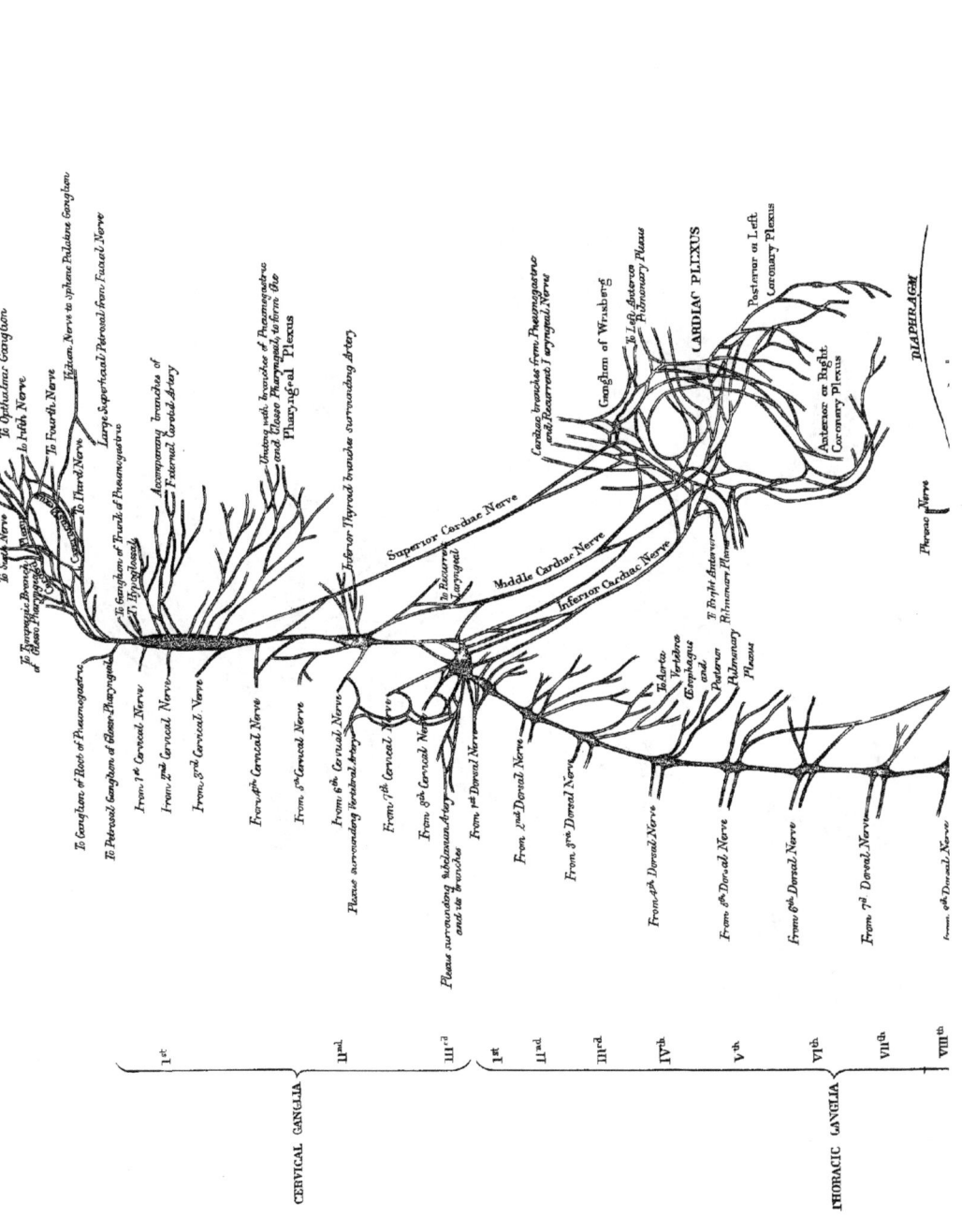

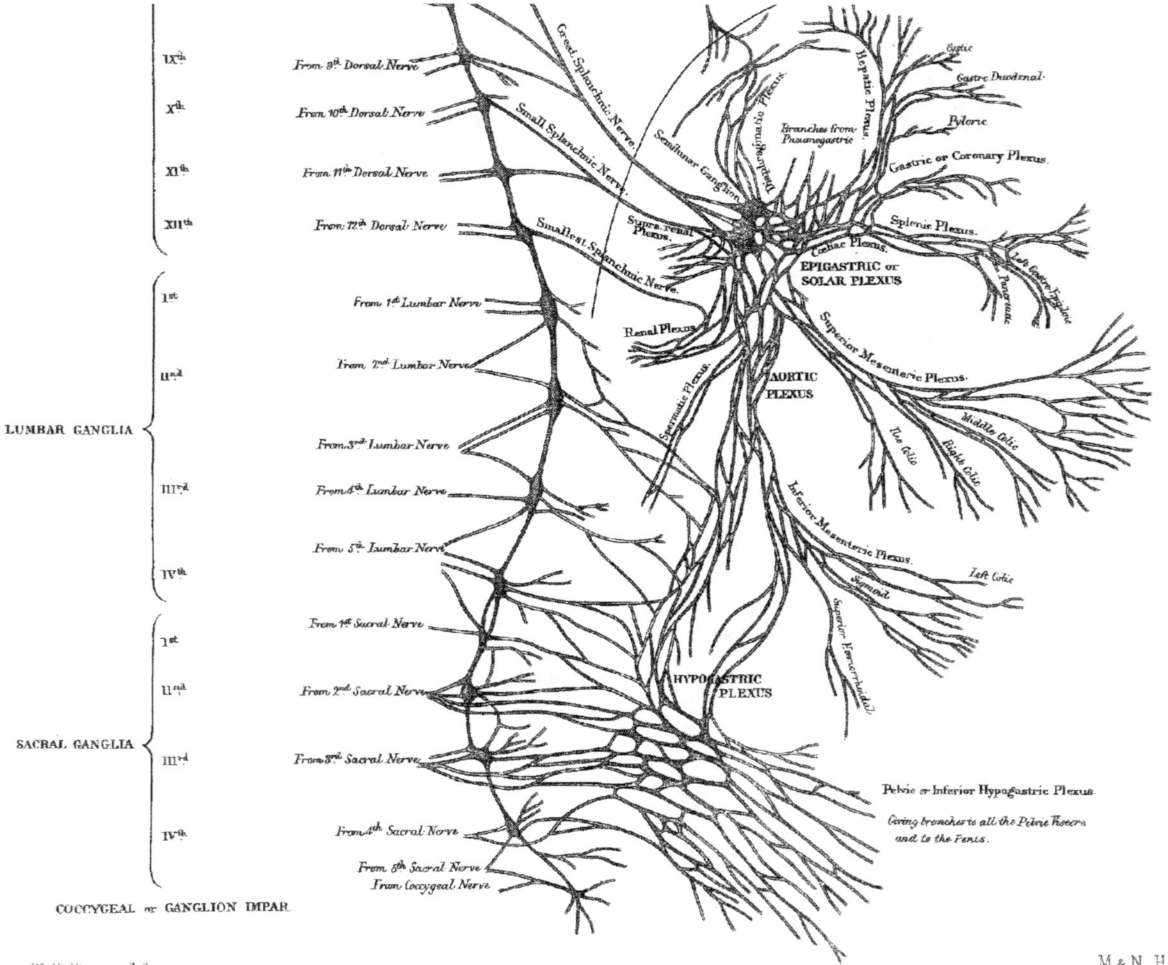

W. H. Flower del.

M & N. Hanhart imp

PLATE V.
THE SYMPATHETIC SYSTEM OF NERVES.

The nerves of this system, also called TRISPLANCHNIC, GANGLIONIC, or NERVOUS SYSTEM OF ORGANIC LIFE, are chiefly destined to supply the viscera. They have abundant communications with the cerebro spinal nerves, in which the fibres of the two systems appear mutually to interchange.

The small ganglia, connected with the fifth cranial nerve, are generally considered as belonging to the sympathetic system. They have been figured in Plate I, and are not repeated here. The ganglia on the glosso-pharyngeal and pneumogastric nerves and those on the posterior roots of the spinal nerves, are also by some anatomists reckoned as part of it.

The remaining, and by far the largest portion (which is illustrated in this plate), consists of two chains of ganglia connected by intervening cords, situated in the posterior part of the neck, thoracic, and abdominal cavities one on each side of the vertebral column and extending from the upper part of the cervical region as far as the coccyx, where they unite in a single small ganglion ($G\ impar$). Each chain usually consists of twenty-four or twenty-five ganglia, having a generally symmetrical disposition on the two sides of the body. Each ganglion is connected above and below with the neighboring ganglion of the chain; externally it has communication with one or more of the spinal nerves, and internally it sends off branches which mostly enter into the formation of certain large plexuses (prevertebral) situated near the median line in the visceral cavities of the body. These, after receiving further accessions from nerves of the cerebro-spinal system, send off branches for distribution to the various organs of the neck thorax, abdomen and pelvis. In these plexuses are situated many ganglia, each of which appears to be a centre for the development or modification of nerve force. This portion of the system, like the parts which it supplies, shows an absence of bilateral symmetry. The branches in reaching their destination almost always accompany bloodvessels, forming a fine network around them.

In the diagram, to avoid useless repetition and obscurity, only one side of such parts of the sympathetic system as are double and symmetrical is given, therefore only one of the ganglionated cords with the ascending cranial branch from the first ganglion appears. The cardiac plexus is, like the organ it supplies single; the cardiac nerves on both sides converging into it. The solar and hypogastric plexuses are also single, and situated in the median line. On each side of the former where the great splanchnic nerve joins it, a large ganglion (semilunar) is placed. Of the secondary plexuses derived from it, the diaphragmatic supra-renal, renal, and spermatic are double, as are the arteries they accompany, but the hepatic, coronary, splenic, superior and inferior mesenteric, and aortic are single and asymmetrical.

The hypogastric plexus divides below into two parts which are situated on either side of the pelvic cavity, and give off the inferior hæmorrhoidal and vesical plexuses, with prostatic and cavernous, or ovarian, vaginal, and uterine branches, according to the sex of the subject. It will be observed that this portion of the system is abundantly reinforced by branches which enter into it directly from the sacral nerves, besides those that pass through the ganglionated cord.

PLATE VI.

DISTRIBUTION OF THE CUTANEOUS NERVES

This diagram is intended to show the sources from which the sensibility of the different regions of the cutaneous surface is derived

The position of the dotted lines which form the boundaries must be regarded only as approximative, as the exact distribution of the cutaneous nerves varies somewhat in different subjects, and as they interlace and communicate freely where they come in contact

In order to trace the surface nerves to their connection with the cerebro-spinal axis with greater facility, an initial reference is given below the name of the branch to the main trunk or plexus from which it proceeds

The explanation of these references is as follows —

    1 V   First (ophthalmic) division of fifth cranial nerve
    2 V   Second (superior maxillary) division of the same
    3 V   Third (inferior maxillary) division of the same
    C P   Cervical Plexus
    B P   Brachial Plexus
    L P   Lumbar Plexus
    S P   Sacral Plexus

Plate VI

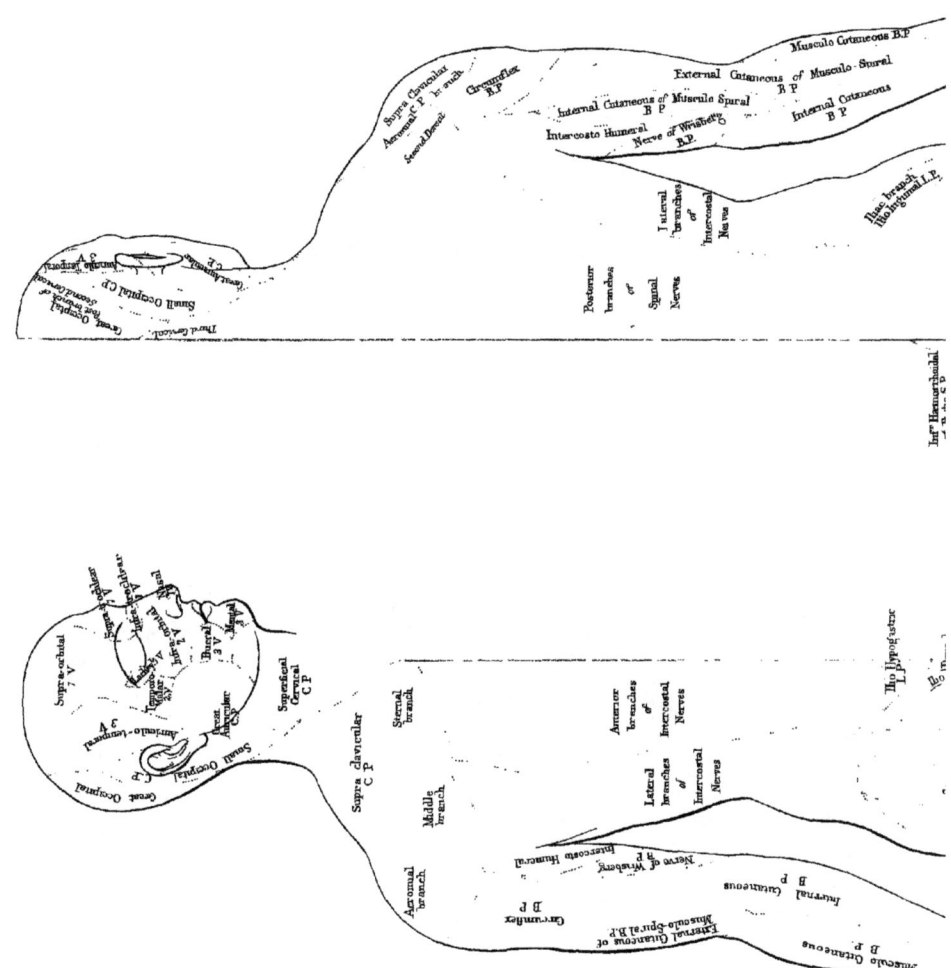

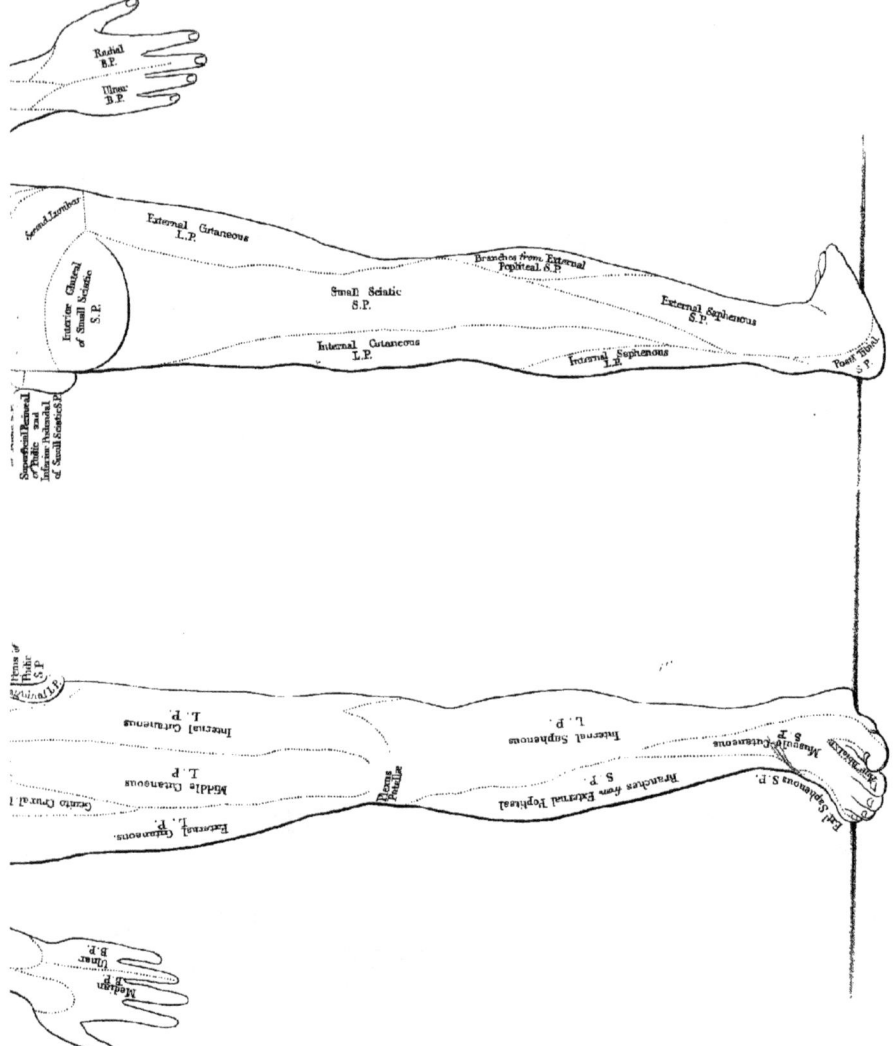

CPSIA information can be obtained
at www.ICGtesting.com
Printed in the USA
LVHW052136040323
740953LV00009B/273